W9-BKA-473

WONDERS
OF THE WORLD

Quicksand

Other books in the Wonders of the World series include:

Gems
Geysers
Icebergs
Mummies

Quicksand

Kris Hirschmann

San Diego • Detroit • New York • San Francisco • Cleveland
New Haven, Conn. • Waterville, Maine • London • Munich

For more information, contact
KidHaven Press
27500 Drake Rd.
Farmington Hills, MI 48331-3535
Or you can visit our Internet site at http://www.gale.com

LIBRARY OF CONGRESS CATALOGING-IN-PUBLICATION DATA

Hirschmann, Kris 1967–
Quicksand / by Kris Hirschmann.
p. cm. — (Wonders of the world)
Includes bibliographical references and index.
Summary: Discusses what quicksand is, how it is formed, its dangers, and how to avert disasters.
ISBN 0-7377-1392-5 (hardback: alk. paper)
1. Quicksand—Juvenile literature. [1. Quicksand.] I. Title. II. Wonders of the world (Kidhaven Press)
QE471.2 .H57 2003
552'.5—dc21

2002010420

Printed in the United States of America

CONTENTS

CHAPTER
ONE

What Is Quicksand?

The word **quicksand** is enough to make some people shiver with fear. This is probably because many movies have presented quicksand as a mysterious substance that sucks people and animals to their deaths. Movie quicksand is usually found in out-of-the-way places where most people never go.

Actual quicksand, though, is very different from the movie kind. First, quicksand is not hard to find. It is common, and it occurs in many different regions and environments. Second, real quicksand is not mysterious. It is a simple substance formed by natural processes that are easy to understand.

Understanding Quicksand

Calling quicksand a substance is a bit misleading. Quicksand is more of a state than a substance. Quicksand is just sand with so much water between its particles that it becomes soupy. When the waterlogged sand can no longer support weight, it is said to be **quick**.

Water's effect on sand can be seen at any seashore. Far from the water's edge, the sand is dry and soft. Although a person's weight shifts the surface layer of the

A man struggles to breathe as he is sucked under by quicksand in a scene from the 1962 movie *Payroll.*

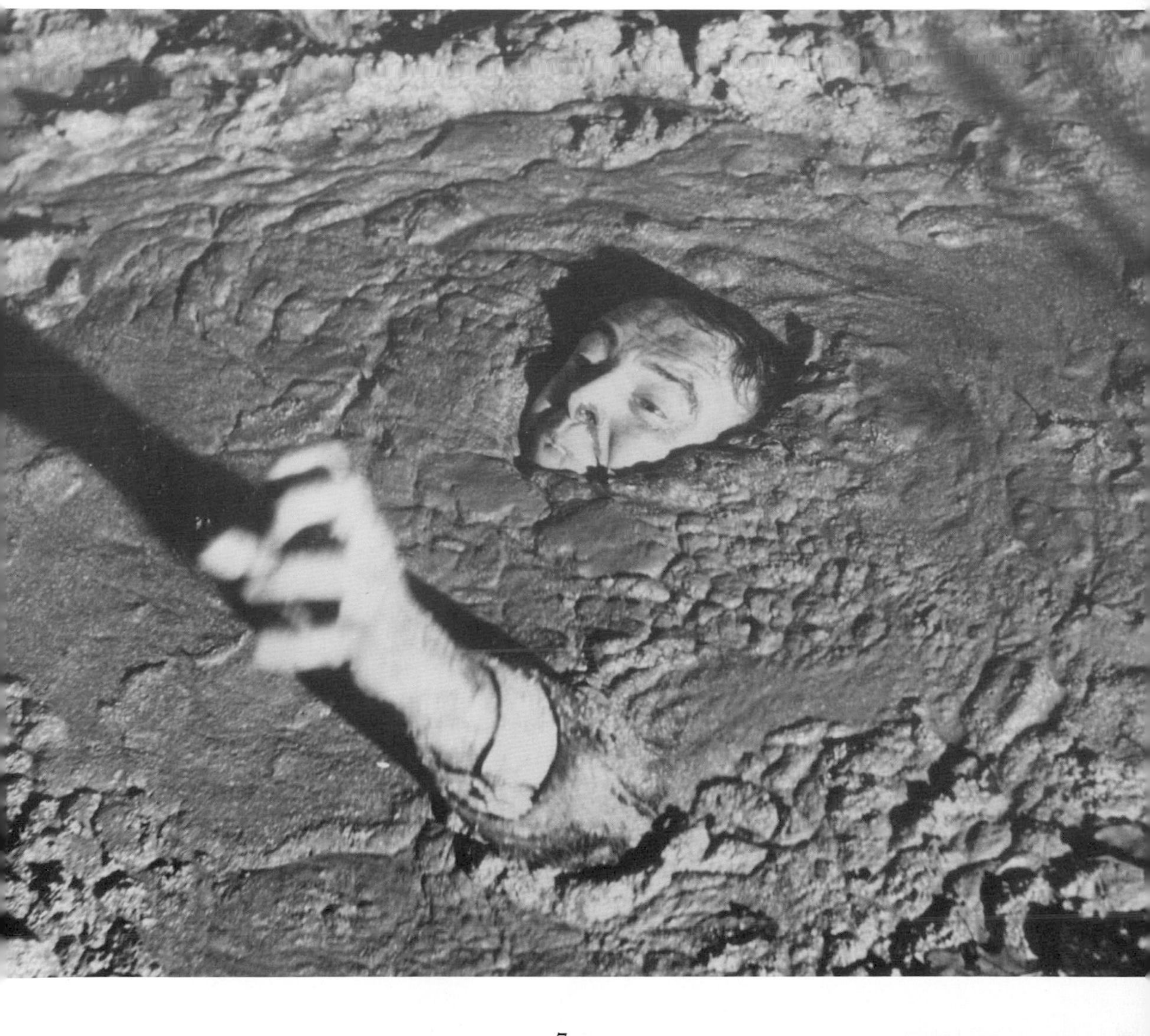

dry sand, there is no danger of sinking. The ground is solid below the top couple of inches.

Farther down the beach but still above the waterline, the sand becomes damp. A little bit of water helps sand particles to stick together, so damp sand is hard-packed and easy to walk on. A walking person barely leaves a dent on this part of the beach.

In the area where waves roll onto the land, the sand changes once again. Here, water sinks into the sand and makes it loose and soft. The sand shifts easily, especially after a wave has washed over it. For a few seconds the surface acts like shallow quicksand. People who stand in the surf and wiggle their feet can make themselves sink a few inches before the quicksand effect fades.

Two children wriggle their toes in the shallow quicksand that forms in the surf.

As its name suggests, quicksand is usually made of sand. But sand is not the only substance that can become quick. Dirt, silt, pebbles, and gravel can also **quicken** under the right conditions.

How Quicksand Forms

The process that creates quicksand is simple. To understand it, though, it is important to first understand the forces acting on substances that are not quick.

Most of the time, sand and other quicksand-forming materials form stable surfaces. These materials are stable because their particles have weight. The weight of the particles pushes downward. This downward push creates a force called **friction.** Friction stops touching objects from sliding against each other. Because of friction, the particles do not move much and they can support the weight of an animal, a person, or even a heavy vehicle.

But conditions change when underground water forces its way between the particles. The water might be bubbling up from a buried spring, or it might be flowing in an underground stream. Whatever the source, the water squeezes between the particles. It separates them just enough to cancel the forces that hold them together. As friction disappears, the once-solid material **liquefies,** which means it starts to behave like a liquid. It is now quicksand. As long as the water continues to flow, the particles will stay in their liquefied state.

Underground water must flow at just the right rate for quicksand to form. If the water flows too slowly, it

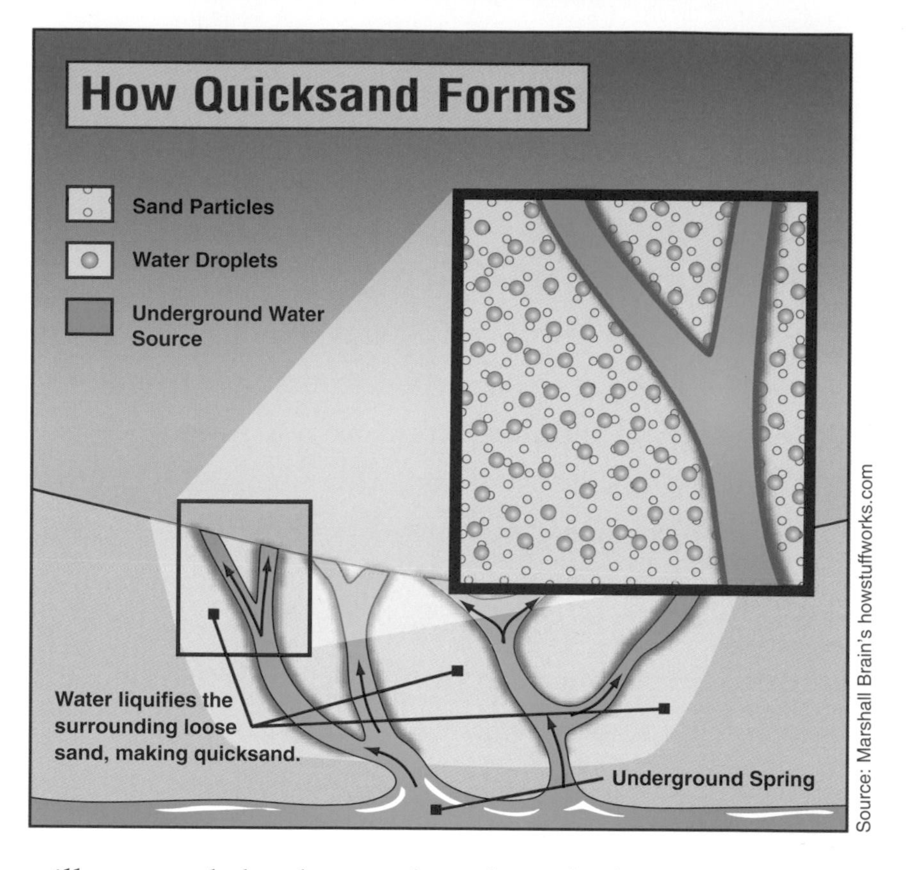

Source: Marshall Brain's howstuffworks.com

will not push hard enough to liquefy the ground. If it flows too quickly, it will wash particles away instead of seeping between them. It takes a moderate but steady stream of water to create quicksand.

Where Quicksand Is Found

Because water is the key ingredient in quicksand, quicksand is usually found in watery places. Marshes, rivers, creeks, and swamps often hide pools of quicksand. The swamps of the Florida Everglades, for example, are full of quicksand, and so are the coastal marshes of South Carolina and Louisiana. Quicksand is also common along seashores on both the Atlantic and Pacific coasts of the United States, especially near sand dunes. Quicksand can also be found throughout rough, hilly country such as the Colorado mountains and the wilds of Texas and Utah. In these areas, underground caves and cracks channel plenty of running water. Where the water rises, quicksand often appears.

Quicksand is not usually found in very flat regions because underground springs seldom form in these areas. Fast-flowing rivers that cut through gorges are usually safe from quicksand, too. The water in these areas moves so quickly that sand and other quicksand-forming materials are washed away. And despite all the sand, deserts almost never contain quicksand. Deserts are dry places and quicksand will not form without water, no matter how much sand exists.

Avoiding Quicksand

Knowing where quicksand is and is not likely to form helps hikers to keep themselves safe. By paying attention and knowing how to recognize quicksand, people who travel through quicksand-prone areas can sometimes spot quicksand before stepping into it.

However, even the most careful hiker might not notice a quicksand pool. Quicksand is often covered by grass, leaves, and sticks. This makes the quicksand hard to spot. Also, the heat of the sun sometimes bakes the top layer of a quicksand pool into a dry crust. The liquefied material below is hidden until something breaks the thin surface layer.

Sometimes quicksand is impossible to see because it is underwater. Underwater quicksand often snags animals and people trying to cross rivers and streams. These dangerous areas cannot be mapped or marked because they change constantly. A riverbank or a marshy hollow may be solid one day and quick the next, depending on water and weather conditions.

Every now and then, a person even creates quicksand where none existed before. For this to happen, a substance must be almost, but not quite, quick. If a person walks onto the substance, it becomes unstable and turns

A young African elephant starves to death after getting stuck in a shallow pool of muddy underwater quicksand.

Quicksand Myths

Fiction	Fact
Quicksand sucks you down.	Anyone who steps into quicksand will sink only to their chest or shoulders. Since quicksand is more dense than water, it is easier to float, and almost impossible to sink completely under.
Quicksand is alive.	Quicksand is not alive. When a person steps into quicksand it moves and ripples with the person's motion which gives the appearance that it is alive.
Leaches and worms live in quicksand.	Quicksand is too wet for worms and too solid for leaches. There is seldom anything alive in quicksand.
Quicksand is bottomless.	There are areas of very deep quicksand, but none that are bottomless. Most quicksand is just a few inches deep to less than waist deep.
Quicksand is found in the desert.	Because it needs a water source, quicksand can be found in swamps, marshes, creek beds, bogs, and on riverbanks and beaches.

to quicksand. The person starts sinking into land that was solid just seconds earlier.

Staying Safe Is Easy

Although quicksand can be hard to avoid, it is easy to understand. Most people who enjoy hiking in quicksand areas have taught themselves about this gooey substance. They know what it is, how and why it forms, and where it is likely to be found. They also know how to avoid it. A little knowledge when walking through quicksand country can help you stay safe from quicksand traps.

CHAPTER
TWO

Caught in Quicksand!

The terrified girl was sinking fast in the oozing quicksand. "Please help me!" she cried, her voice cracking with fear. Chest-deep in muck, the girl stretched an arm toward a friend who was standing on the edge of the quicksand pool. But it was too late. The quicksand pulled its victim under with lightning speed. Within moments, the unlucky girl had disappeared. Ripples on the quicksand's surface were the only sign of the struggle that had just occurred.

Frightening scenes such as this one are all too common in the movies, in books, and on television shows. In the real world, however, quicksand encounters are not usually so dramatic. In most cases it is rare for quicksand to suck a person completely underneath.

A terrified woman trapped in quicksand screams for help. In the movies, quicksand is usually portrayed as being more dangerous than it really is in nature.

But although real-life quicksand is not as deadly as the fictional kind, it is still dangerous. Quicksand does trap people. In certain circumstances, it can even kill.

The Facts About Falling In

Many people fear quicksand because they think it will pull them underneath. But this is not likely to happen. One reason for this is that quicksand pools are not usually very deep. Most of the time a person sinks only a few inches before hitting bottom. Some pools are a few feet deep, but this is no problem for most people. An adult's feet will land on solid ground by the time the quicksand reaches his or her waist.

Large, deep quicksand pools do sometimes form. But even a deep pool will not usually swallow a person, for one simple reason: Quicksand is denser than the human body. This means that quicksand is heavier and thicker than human flesh. As a result, the upward push of quicksand is greater than the downward push of a person. So unless a person is weighted down, he or she will always float in quicksand.

Before floating, though, a person will sink quite a bit. The speed of sinking depends on the **density** of the quicksand. A person will sink quickly in loose, runny quicksand, which may be only a little bit denser than the human body. Sinking will be slower in thick quicksand, which is much denser and therefore supports more weight.

After sinking up to his chest in quicksand, a man floats to stay alive, in this scene from the movie *The Devil at 4 O'Clock.*

Whether quicksand is thick or runny, a person usually sinks to chest level before floating. The lungs inside the chest act like giant balloons to keep the quicksand victim from going under.

Sinking Without a Trace

One special circumstance exists in which a person can sink beneath the surface of a quicksand pool. A person who is carrying a backpack or another heavy object may be denser, on average, than quicksand. If the person cannot remove the heavy object, he or she may be completely sucked into a pool of quicksand.

According to one account, this happened to a hiker named Jack in 1964. Jack was hiking with a friend in the swamps south of Florida's Lake Okeechobee when he stepped into quicksand. The quicksand was especially loose and watery, and Jack immediately sank to his chest. With his arms caught in the muck, Jack could not reach the release hooks on his backpack. Jack struggled to free himself as the heavy pack dragged him deeper and deeper into the muck. But he could not shake the backpack loose. Before long, Jack sank beneath the surface of the quicksand and drowned.

A similar fate may have befallen several soldiers in long-ago France. Legend has it that a band of armored men wandered into quicksand when crossing a boggy area. The soldiers could not remove their heavy armor in time to save themselves. They all died after sinking below the quicksand's surface.

A harmless drive on the beach in Texas turns into a disaster when this woman's car gets stuck in quicksand.

People are not the only quicksand victims. Cars, trucks, and other vehicles are much denser than quicksand, so they will sink if a quicksand pool is deep enough. In 1945, for instance, a U.S. Army driver lost his truck and all its cargo in a German quicksand bog. And in the late 1800s, a Kansas Pacific Railroad train was sucked into quicksand in Colorado after falling into a flooded creek bed. Most of the train was recovered. But even though workers dug down fifty feet, they never found the train's two-hundred-ton engine. It had disappeared without a trace.

Stuck Tight

Although sinking completely into quicksand can happen, it is very rare. A bigger danger is simply getting stuck beyond any hope of escape. Once stuck, a quicksand victim may die from exposure or starvation if help does not arrive.

People and animals get stuck in quicksand for two main reasons. One is a lack of anything to push or pull against. The other is resistance from the quicksand itself. Moving an arm or a leg in quicksand can create a **vacuum** (an absence of air). Vacuums resist movement and may keep a quicksand victim's limbs stuck firmly in place.

Some evidence exists that dinosaurs died after becoming stuck in quicksand. Many dinosaur remains have been found in fossilized quicksand beds in Mongolia. Scientists believe that these remains were left when dinosaurs were trapped in quicksand bogs and died of starvation. One particularly dramatic fossil, found in the 1960s, shows two dinosaur skeletons locked in battle. It appears that the dinosaurs were fighting when they fell into quicksand. Trapped and unable to escape, they soon died.

Humans, too, have died after being trapped in quicksand. According to one story, such a death occurred in the early 1900s in Arkansas. A hunting party was hiking along a river when suddenly it came upon a man's head on the ground. On closer inspection, the group realized that the head was attached to a body—but the body was underground. The man had sunk into quicksand up to the neck, then starved to death when he could not escape.

Even worse fates have befallen some unlucky quicksand victims. In Spain, trapped travelers have been

Fossilized dinosaur tracks can be seen in a dried-up creek bed. Scientists believe that creek beds like this once held quicksand traps.

partly eaten by vultures. And in Alaska, hunters have been mauled by bears after being caught in quicksand. Stuck people have no way to defend themselves, so they are easy pickings for hungry animals.

Drowning Danger

Getting stuck in quicksand can be deadly in another way. It may lead to drowning if the quicksand is in a watery area.

Underwater quicksand is especially dangerous. People sometimes get pulled into quicksand on the bottoms of creeks and ponds, then drown when they cannot keep

A man struggles to pull his friend from a pool of quicksand created by the changing tides at Goodwin Sands in England.

their heads above the water's surface. In 1951 a six-year-old Texas boy died this way. The child was wading in a creek when he got caught in a pool of underwater quicksand. Although the boy sank only to his waist in the quicksand, the water above the muck was more than three feet deep. Stuck tight with his head below the waterline, the boy soon drowned.

Tidal Areas

Quicksand in tidal areas can also be deadly. If someone is caught in a marshy seaside bog while the tide is out, he or she is in serious danger of drowning when the tide comes back in.

In one dramatic story, a Georgia man got stuck up to his neck in seaside quicksand. The man yelled for hours before someone heard him and called for help. Rescuers soon arrived on the scene, but it was dark and the tide was coming in. The water rose higher and higher as the frantic rescuers tried to find the quicksand victim. They finally found their man, who was bending his neck backward to hold his nostrils out of the water. Another few minutes and he would have been dead.

This Georgia man was luckier than a young woman named Adeana, who became stuck in an Alaskan bog in 1988. Adeana's husband and several friends worked hard to dig her out of the quicksand as the tide swept in, but they could not pull her loose. Someone gave Adeana a snorkel so she could breathe as the water rolled over her head, and the digging continued. But nothing worked.

Masked by plants and grasses, quicksand may form in the pools of marshy bogs.

The quicksand would not release Adeana, and she drowned during the rescue effort.

A Grim Reminder

Deaths from starvation, drowning, and other quicksand-related causes are gruesome and frightening. Luckily, they are also very rare. Quicksand is not usually deadly or even very dangerous. But tragedies do occur. Each disaster is a reminder that quicksand must be taken seriously. For the unprepared or the unlucky, it can be a death trap.

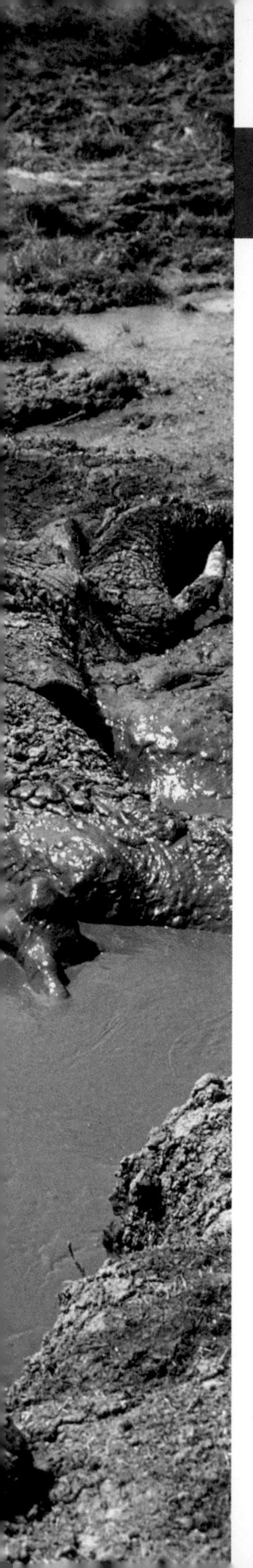

CHAPTER
THREE

Incredible Escapes

By all accounts, falling into quicksand is a frightening experience. People who find that the ground has turned to liquid beneath their feet almost always feel a moment of panic. But it is better to remain calm. There are many ways for people to get themselves out of quicksand pools. And even if a person cannot escape without help, rescuers can usually pull quicksand victims to safety.

In most cases getting out of quicksand is fast and easy. Once in a while, though, it is not as simple for a quicksand victim to get loose. Some people have had to make extraordinary efforts to escape from quicksand pools on their own. Others have been saved by difficult and heroic rescue operations. These incredible quicksand escapes make great stories. They also add to quicksand's reputation as something to be feared and avoided.

Quicksand on the Trail

Some of the earliest quicksand escape stories come from the days of the cowboys. Quicksand was a constant danger in the Old West, where many rivers had quicksand bottoms. Cowboys often drove herds of cattle across these dangerous rivers, and they never knew when a cow might suddenly find itself sinking into the muck.

According to cowboy **lore,** cattle nearly always panicked and struggled when they started to sink. Struggling increases the rate of sinking, so caught cows tended to go

As cattle cross a riverbed they may sink if quicksand lurks below.

down fast. Their movements drove them deeper and deeper into the muck. Without help, a cow stuck in quicksand was almost sure to die.

To rescue a cow, a cowboy stripped off his clothes, then jumped into the quicksand alongside the animal. He worked quickly to free the cow's front legs and loosen the hind legs. Next the cowboy tied a rope around the cow's horns, freed himself from the quicksand, and tied the other end of the rope to his horse's saddle. With a little urging and a little luck, the horse then hauled the frightened cow out of the quicksand.

Once a cow was free, it was safe from danger. For the cowboy, however, the most dangerous part of the rescue was still to come. A rescued cow was usually terrified and angry, and it often charged at the man who had just saved its life. To avoid being gored, an experienced cowboy always cut his rescue rope and rode away the instant a cow was safe.

Swimming to Safety

It might seem that cattle-rescuing cowboys were risking their lives when they entered quicksand. But cowboys were not afraid of quicksand because they knew one important trick: People can swim in quicksand if they move slowly enough. As long as a person does not panic, he or she can paddle to the edge of even the largest quicksand pool.

One Louisiana man escaped from quicksand in this way. The man was walking in a marsh when suddenly the ground became soft under his feet. Without real-

A duck hunter struggles to free himself from quicksand hidden in this Louisiana marsh.

izing it, the man had stepped onto an unstable surface, and his weight was quickening the ground. Before the man could step back, the land around him had turned into a gigantic pool of quicksand. Right away the man sank to his chest—but he did not panic. Very slowly, he made swimming motions with his arms and legs.

After nearly an hour of careful swimming, the man reached the edge of the quicksand pool and climbed out unharmed.

A scientist with the U.S. Geological Survey had a similar experience in the early 1900s. This scientist was exploring a river in Colorado when he became stuck in a quicksand bed. He immediately started swimming. But the muck was thick, so each movement was painfully slow. The scientist did manage to move forward slowly, and he eventually reached the edge of the quicksand. But it took him eight hours to travel a distance of just ten feet. Patience and persistence saved this quicksand victim's life.

Pulled from the Muck

Although swimming is a good way to get out of quicksand, it can be slow. A person can escape much more easily and quickly by pulling or pushing against a solid, fixed object, if one is within reach. Branches, rocks, and even long grass can be lifelines for someone who is caught in quicksand.

Even a piece of fishing line can be enough to save a quicksand victim's life. One New York fisherman discovered this the hard way when he stumbled into quicksand near Long Island Sound. Trapped up to his hips but still holding his fishing rod, the man cast his line toward a large tree branch about twenty feet away. He eventually snagged the branch and slowly began reeling it in. The delicate fishing line could have broken under the strain, but the fisherman was lucky; the line held. Soon the fish-

Exhausted from treading the soupy sand, this woman is finally pulled to solid ground in the 1950 movie *Captive Girl.*

erman was holding the branch in his hands. He used it to pry his legs out of the quicksand. Before long the man had escaped from the sucking ooze.

In another survival story, a twelve-year-old New Jersey boy named Khristopher used his own body to rescue a trapped friend. The friend had sunk up to his neck in quicksand. There were no adults around to help, and there were no solid objects that might be used to help the friend escape. So Khristopher wrapped his hands

Fishermen who get caught in quicksand can snag solid objects with their fishing lines and reel themselves to safety.

around some tree roots and flung his legs into the quicksand. The stuck friend seized Khristopher's feet and held on tight. Khristopher was then able to pull both himself and his friend to safety.

Helicopters to the Rescue

Rescuers have many ways of pulling quicksand victims out of the muck. If a stuck person is close to solid land, a rescuer may be able to pull him or her out by the hands. Rescuers may also toss a rope to a stuck person, then haul him or her out inch by inch. Or they may lay ladders and boards on top of the quicksand to give the victim something to cling to.

If a quicksand victim is in an especially awkward place, a rescue helicopter may be called in. The helicopter hovers above the stuck person and lowers a rope. The quicksand victim wraps the rope around his or her chest, then is lifted to safety.

The helicopter method works well in most cases. Sometimes, though, vacuum forces in the quicksand hold people so tightly that hauling them out is nearly impossible. If a helicopter pulls too hard, it can break a quicksand victim's bones or even tear the person in half.

This was almost the fate of a man named Tony who became trapped in an Alaskan quicksand bog in 1981. Tony had sunk to his waist in quicksand, which had then turned into thick mud. He was stuck tight. A rescue helicopter soon arrived on the scene and lowered a harness, which Tony slipped around his chest. The helicopter began to rise. But immediately Tony's eyes became wide with pain, and he signaled frantically for the helicopter to stop. He knew that the suction of the mud was too strong. If the helicopter kept pulling, Tony might be torn to pieces.

The helicopter pilot had two big problems. He had to keep careful control of his helicopter, because any sudden movement could seriously hurt Tony. But the

Helicopters are used to rescue people trapped in quicksand. A rope is lowered from the helicopter and a harness is attached before a person is pulled free.

tide was coming in, and Tony would drown if he were not rescued soon. So the pilot had to keep trying. To keep the helicopter from being blown around by the wind, the pilot descended until he was hovering about ten feet above Tony's head. Then the pilot pulled upward with a gentle, steady pressure. After about thirty minutes, the muck finally loosened and a grateful Tony was lifted to safety.

Everyday Escapes

Most quicksand rescues are not as dramatic as Tony's. When people fall into quicksand, they almost always get themselves out without much trouble. And those who do get stuck are usually easy to rescue.

Hikers and other people who travel through quicksand areas can keep themselves safe by understanding quicksand. They can learn the tricks for getting out of quicksand, such as swimming and pulling on solid objects. If a person knows what to do and does not panic, he or she has every chance of escaping from quicksand.

CHAPTER
FOUR

When Earth Turns to Liquid

Quicksand is usually found in shallow pools above underground water sources. Although the pools are not permanent, they do last as long as the water keeps flowing.

Occasionally, though, solid earth can dissolve into quicksand even without an underground water source. The process that creates this type of quicksand is called **liquefaction**, and it occurs when earthquakes shake the ground. Quicksand created through liquefaction is temporary. It may last just a few seconds before the earth hardens again.

How Liquefaction Happens

Liquefaction occurs only in **saturated** soils, meaning that the spaces between the soil particles are completely filled with water. Although the soil is solid, it is damp throughout.

Under normal conditions, the particles in damp soil push against each other because of their weight. The pushing creates friction forces that hold the particles together and make the soil solid. When an earthquake shakes the ground, however, the forces change. Vibrations make the soil particles push even harder as they try to squeeze together. But the particles cannot get closer because of all the water between them. As the particles push harder and harder, the water pressure in the ground rises and rises. If the water pressure becomes high enough, it breaks the friction forces that hold the particles together, and the soil turns into quicksand.

A man surveys the ground that had been liquefied by an earthquake.

Quicksand created by earthquakes does not last long. The soil stays quick only as long as the ground is shaking. When an earthquake ends, the soil particles stop pushing against each other. As a result, the water pressure drops and the soil becomes solid once again.

The Great San Francisco Earthquake

Often only a small amount of soil liquefies during an earthquake. Sometimes, however, an earthquake creates so much quicksand that buildings and other structures may be damaged. This is called **ground failure**.

Ground failure occurred in a 1906 earthquake. This mighty earthquake struck California early in the morning on April 18 and began to shake the land. Buildings toppled up and down the West Coast, from Los Angeles to Oregon. But the city of San Francisco, which was built on the water's edge, was shaken the hardest of all. As San Francisco shook, the damp earth on which the city was built began to turn into quicksand. Under-

On April 18, 1906 an earthquake demolished San Francisco. The city's damp foundation turned to quicksand, which caused underground pipelines to snap.

Liquefaction caused an enormous landslide during the 1964 Alaska earthquake.

ground pipelines all over the city snapped as the earth around them became liquid.

The great earthquake lasted less than a minute. When it was over, San Francisco was ablaze with fires that had started during the quake. Firefighters rushed to put out the fires, but the underground pipes that carried water through the city were broken. Without any water to fight the blazes, there was little firefighters could do. They watched helplessly for the next four days as San Francisco burned. By the time the fire finally exhausted itself, the city had been destroyed.

Landslides in Alaska

Ground failure caused another spectacular disaster in 1964, when a strong quake hit the southern coast of Alaska. The quake struck Prince William Sound on

An aerial view of Valdez, Alaska shows the damage caused by a 1964 earthquake.

March 27 and shook the ground for more than three minutes. During this time, a mile-long strip of seaside land liquefied. Because the strip was on a hill, it immediately started sliding downward. By the time the landslide ended, seventy-five buildings that had stood on the once-solid land had been washed into the ocean. Many of the buildings' residents were crushed, drowned, or buried in quicksand during the process.

This landslide was not the only quicksand-related disaster that occurred during the 1964 Alaska earthquake. Many roads crumbled when the solid ground beneath them turned to liquid. Also, more than two hundred bridges were damaged or destroyed by liquefaction during the quake. A lot of the land near rivers turned to quicksand and began to spread. This spreading shifted the foundations of bridges and either snapped or squeezed the roadways above.

Destruction in Japan

Just a few months after the Alaska disaster, another earthquake caused severe liquefaction in the Japanese town of Niigata. As the ground shook on June 16, 1964, the earth throughout this port town started to liquefy. Bridges, highways, dock areas, oil refineries, and railroads all over Niigata were damaged as the land turned to quicksand. Underground sewage tanks floated to the surface. Cameras captured scenes of the ground actually boiling as liquid soil bubbled its way into open air.

The most dramatic problem occurred near the Shinano River, where the ground failed under the Kawagishi-cho apartment complex. Several of the complex's four-story buildings tipped over as their foundations turned to liquid. But because the tipping happened slowly, the structures

Buildings in Niigata, Japan lean and fall as the ground beneath them turns to quicksand.

The 1964 Niigata earthquake caused widespread damage including liquefaction and flooding.

were not badly damaged. Many residents of one especially tilted building escaped by climbing out of their windows and walking down the structure's nearly level face.

In another part of town, the Niigata airport terminal building sank nearly four feet into a giant quicksand pool when the ground beneath it liquefied. The terminal could not be repaired.

How Likely Is Liquefaction?

Earthquakes such as the ones that devastated San Francisco, Alaska, and Niigata show that liquefaction can cause a lot of damage. For this reason, scientists are working to identify places where liquefaction is likely to occur. They have created liquefaction maps that pinpoint danger areas—areas of damp soil in earthquake zones. Using liquefaction maps, builders may decide to use extra-strong construction techniques on new buildings. Before building, they may also prepare the soil in ways that make the ground less likely to liquefy. Both of these techniques can help reduce the damage if liquefaction occurs.

But liquefaction maps are based only on probability, not on fact. No one really knows where liquefaction will

A plane speeds down a runway that had been damaged by liquefaction caused by an earthquake.

happen because no one knows exactly where earthquakes will strike. For this reason, many builders do not take the steps that would keep a building safe from liquefaction. Sea coasts, bay shores, and other danger areas are jammed with unsafe buildings that could sink into the soil if an earthquake hit the area. And these buildings are full of people. If a building were destroyed because of liquefaction, these people could lose their belongings or even their lives.

Although it is impossible to predict earthquakes or liquefaction, one fact is certain. Sometime, somewhere, a major earthquake *will* strike, and the soil *will* liquefy. No one can guess how much damage will be done or how many people will die. But if the earthquake strikes in a heavily populated area, hundreds or even thousands of people could lose their lives. Such a disaster would certainly be remembered as the greatest quicksand catastrophe of all time.

Glossary

density: The amount of matter packed into a certain space.

friction: A force that stops objects from sliding against each other.

ground failure: When enough land liquefies to damage structures.

liquefaction: The process that turns a solid into a liquid.

liquefies: The process of becoming liquid.

lore: Traditional knowledge; legend.

quick: Moving, flowing; unable to support weight.

quicken: To become quick.

quicksand: Sand, silt, or other material that is saturated with water and cannot support weight.

saturated: Holding the maximum possible amount of water.

vacuum: The absence of any matter, including air.

For Further Exploration

Books

Judy Donnelly and Sydelle Kramer, *Survive! Could You? Quicksand, Earthquake, Shark Attack, and More.* New York: Random House, 1993. This book presents many life-or-death situations (including being trapped in quicksand) and explains how to survive each one.

Alison Lester, *The Quicksand Pony.* Boston: Houghton Mifflin, 1998. This fictional story starts when Bella the pony is trapped in quicksand and is left behind by her owners.

Janet Nuzum Myers, *Strange Stuff: True Stories of Odd Places and Things.* North Haven, CT: Linnet Books, 1999. Read about quicksand, Bigfoot, zombies, the Bermuda Triangle, and other bizarre things in this book.

Websites

How to Make a Box of Quicksand (www.hunkins experiments.com). Step-by-step written instructions and illustrations explain how to make quicksand in a box.

How Quicksand Works (www.howstuffworks.com). Simple text and diagrams explain how quicksand works. Follow the links at the top of the quicksand page to learn more, including how quicksand forms and how to escape.

The Quicksand Page (www.dellamente.com). This site has a lot of good information about quicksand seen in the movies, on television, and in books. Includes video clips.

The Soil Liquefaction Web Site (www.ce.washington.edu). Follow the links on the introductory page to learn all about soil liquefaction.

Index

Picture Credits

Cover Photo: © Peter Johnson/CORBIS
© Tom Bean/CORBIS, 20
© Niall Benvie/CORBIS, 23
© Bettmann/CORBIS, 40
© Lloyd Cluff/CORBIS, 35, 37, 38, 41
© CORBIS, 36, 39
© Corel Corporation, 30, 32
© Philip Gould/CORBIS, 27
© Hulton Archive by Getty Images, 21
© Peter Johnson/CORBIS, 12
© Robert Maass/CORBIS, 25
Brandy Noon, 10
Jeff Paris, 13
Photofest, 7, 15, 16, 29
© Vince Streano/CORBIS, 18
© David Turnley/CORBIS, 8

About the Author

Kris Hirschmann has written more than sixty books for children, mostly on science and nature topics. She is president of The Wordshop, a business that provides a wide variety of writing and editorial services. She holds a bachelor's degree in psychology from Dartmouth College in Hanover, New Hampshire. Hirschmann lives just outside of Orlando, Florida, with her husband, Michael.